INTRODUCTION

Frank Schaffer's Algebra for Everyday is filled with a lot of handy, stimulating activities students can complete to learn some valuable algebra skills. These relevant, interesting activities make algebra more meaningful for students, and they also help students see the important role algebra plays in their lives every day.

Easy to use, these reproducible activities will help students learn such algebra concepts and skills as logic, visual discrimination, problem solving, prime and composite numbers, patterning, prime factorization, solving equations, area of polygons, Pythagorean theorem, and much more.

Students will enjoy completing puzzles, filling in charts, designing cubes, creating patterns, drawing figures, and more as they learn the many concepts involved in algebra.

Included with some of the activities are game pieces or other similar manipulatives. To use, simply cut them out, laminate them, and store them in resealable bags so that they can be used over and over. Also included is an answer key on pages 75–79 to make checking students' work simple.

Regardless of students' strengths or weaknesses when it comes to working with algebra concepts, this book provides the right variety of activities that will allow every student to experience success. You will be pleased as students begin grasping important algebra concepts and developing valuable algebra skills as they have fun working interesting and challenging activities.

everyday **algebra**

PUZZLED?

Fill in the blanks to correctly total each row and column based on the given operations.
For more fun—practice the first puzzle on your own, then race with a friend to finish the second!

1. Practice

3	x		÷		=	6
+	■	+	■	x	■	−
	÷		+		=	
−	■	−	■	−	■	+
	+		−		=	5
=	■	=	■	=	■	=
2	x	5	−		=	5

2. Race

5	x		−		=	6
+	■	x	■	+	■	÷
	÷		+		=	3
−	■	−	■	−	■	+
	x		−		=	
=	■	=	■	=	■	=
6	−		+	5	=	9

FS-10606 Everyday Algebra

ROMAN TOOTHPICKS

Arrange the toothpicks in each problem below to make each statement true.

1. Move two toothpicks.

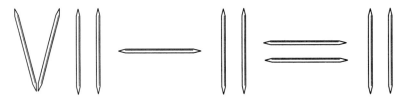

2. Move only one toothpick. (**Hint:** square root)

3. Move two toothpicks.

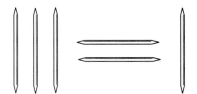

FS-10606 Everyday Algebra

e v e r y d a y   a l g e b r a

COLOR THE CUBE

Small cubes have been stacked and glued together to form this larger cube below.
Use the cube to answer the questions.

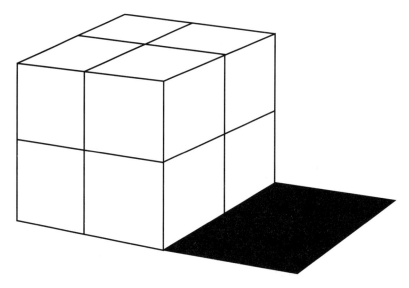

1. How many small cubes were used? _____

2. If the cube was dropped in a pail of paint:

    a. how many small cubes would have paint on three sides? _____

    b. how many small cubes would have paint only on two sides? _____

    c. how many small cubes would have paint only on one side? _____

    d. how many small cubes would not have paint on any side? _____

3. Answer questions 1 and 2 for a 3 × 3 cube.   **1.** _____   **2. a.** _____

    b. _____

    c. _____

    d. _____

4. Answer questions 1 and 2 for a 4 × 4 cube.   **1.** _____   **2. a.** _____

    b. _____

    c. _____

    d. _____

CLEAN UP LOGIC

Mr. Jones reminds his five children that every Saturday they must clean the house before they play. Each child must pick a room to clean. Use the clues below to determine which room each child picks to clean.

Children: Anna, Beth, Christopher, Dean, and Ellen

Rooms: the living room, den, kitchen, bathroom, and bedroom (**Hint:** Fill in the chart below to orgainze the information.)

1. Anna dislikes cleaning sinks.

2. Christopher doesn't make beds.

3. Beth and Dean always turn the cushions when they clean.

4. The kitchen and the den are close. Dean and Ellen enjoy talking while they clean.

5. There are couches in the den and the living room.

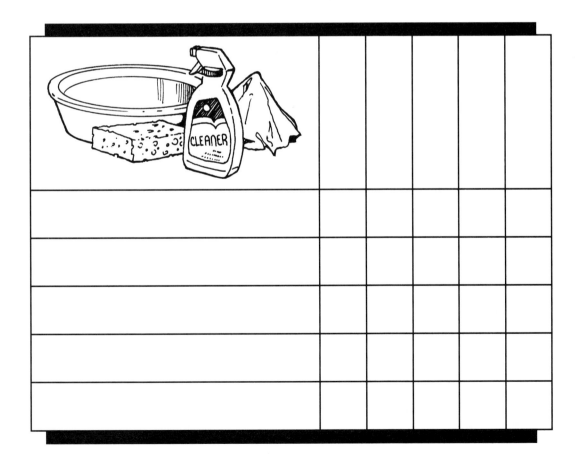

	Anna	Beth	Christopher	Dean	Ellen

_____ _____ _____ _____ _____

SEPARATION ANXIETY

1. Use three straight lines to separate each number in the puzzle below.

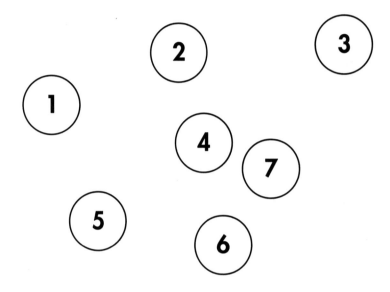

2. Use two straight lines to separate the clock such that the sums of the numbers in each section are equal. (**Hint:** Each section equals 26.)

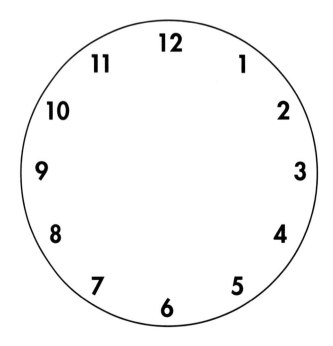

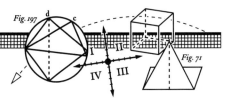

everyday
algebra

BLOCK TO THE BASICS

1. The four blocks below are identical, and one side of each block has no marking at all. In the boxes, draw exactly what the "block" looks like.

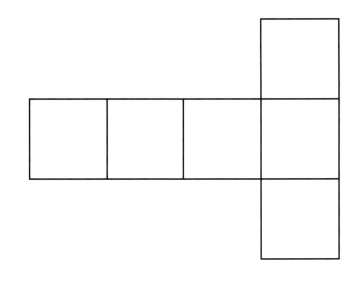

2. Arrange the three blocks below to form a three-digit number that can be divided by seven evenly.

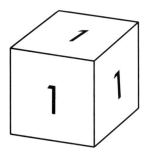

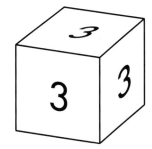

MONEY MADNESS

1. A rancher buys 100 live animals for $100. Chickens cost 50¢ each, goats cost $3.50 each, and cows cost $10 each. How many of each did the rancher buy? Is more than one combination possible?

2. You have 50 coins which have a total value of $1.00. How many coins of each do you have? Is more than one combination possible?

THE THREES HAVE IT

Make the numbers 0–10 using five 3s. (Try to get as many as you can in three minutes. Or race the class to see who finishes first.)

Example:

$$(3 - 3) \times (3 \times 3 + 3)$$
_____ = 0

1. _____ = 1

2. _____ = 2

3. _____ = 3

4. _____ = 4

5. _____ = 5

6. _____ = 6

7. _____ = 7

8. _____ = 8

9. _____ = 9

10. _____ = 10

Name _____ Date _____

WHAT COMES NEXT?

Find the next term in each pattern. Then explain your answer.

1.

A	B	C
D	E	F
G	H	I

,

D	A	B
G	E	C
H	I	F

,

G	D	A
H	E	B
I	F	C

,

_____

_____

_____

_____

2.

 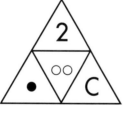 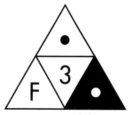 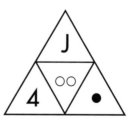 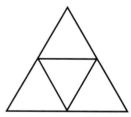

_____

_____

_____

_____

_____

COMPARING VISUAL PATTERNS

The first two figures in each problem demonstrate the change of a pattern.
Draw the figure that simulates that same change with the second set of figures.

Example:

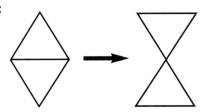 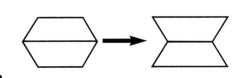

1.

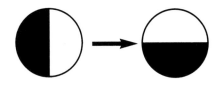 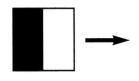

2.

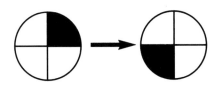 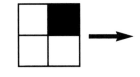

3.

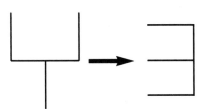 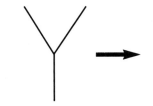

4.

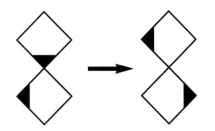 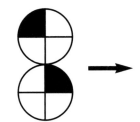

FS-10606 Everyday Algebra

BASKETBALL BONKERS

The Sharp Shooters basketball team scored 38 points in the first half of the Tournament Championship. How many different combinations of baskets did the team make (3 points, 2 points, free throw—1 point)? (**Hint:** Use a table to organize the information in this problem.)

Type of Basket	Quantity					
3 pts.	12					
2 pts.	0					
1 pt.	2					
Total	38	38	38	38	38	38

The Sharp Shooters' opponents, the Brick Bombers, went into halftime 13 points down. What was their score and how many types of baskets did they make?

Type of Basket	Quantity					
3 pts.						
2 pts.						
1 pt						
Total						

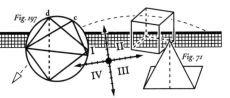

everyday
algebra

TO CUBE
OR NOT TO CUBE?

"Visualize" the figures below to determine if each one will form a cube. Then cut them out and fold them to check your answers.

1.　　　　　　　　　　　　　　　　**2.**

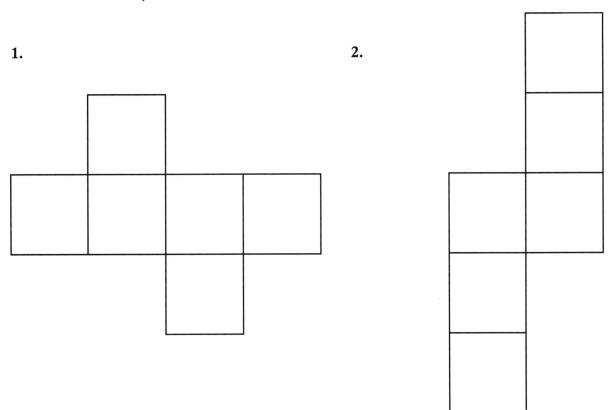

Create two "designs" of your own that can each be formed into a cube.

Name_____ Date _____

PICASSO 101

Practice sketching the figures below.

Find something "geometric" in your classroom and sketch it on the back of this page.

1. cube	**2.** 3-D triangle
3. pyramid	**4.** cone

MATHO THE GREAT

1. I am thinking of three 4-digit numbers. Each number has four different digits. Also in each number, the first two digits, middle two digits, and last two digits are perfect squares. What are the three numbers?

2. I am thinking of three numbers. One number contains 2 digits, one 3 digits, and one 5 digits. The numbers 0–9 are each used only once in all three numbers. If I label these three numbers $x, y,$ and z, then $y = 22x$ and $x \cdot y = z$.

What are the three numbers?

        x y z

____ ____   ____ ____ ____   ____ ____ ____ ____ ____

everyday **algebra**

COUNTRY COUNTIN'

A. Rancher Roy's field has 15 horses and chickens in it. He counts 48 legs. How many horses and how many chickens does Rancher Roy have? Remember, horses have 4 legs and chickens have 2 legs. (**Hint:** Make a chart to organize your thoughts.)

Example:

horses	1	2
chickens	14	13
total number of legs	32	34

B. Farmer Phil sent his farm hand to town to get five sacks of grain. The store manager loaded the grain in his truck and said, "Tell Farmer Phil that sack 1 and sack 2 together weigh 48 pounds; sack 2 and sack 3 together weigh 54 pounds; sack 3 and sack 4 together weigh 46 pounds; sack 4 and sack 5 together weigh 32 pounds; and sack 1, sack 3, and sack 5 together weigh 64 pounds." How much total grain did the farm hand bring back to Farmer Phil, and what did each sack weigh?

WATCH THE CLOCK

1. At which times do the digits on a digital clock contain three identical consecutive numbers? (**Hint:** There are 17.)

Example: 1:11

_____   _____   _____

_____   _____   _____

_____   _____   _____

_____   _____   _____

_____   _____   _____

_____   _____

2. At which times do the digits on a digital clock read the same forward and backward? (**Hint:** There are a lot—list only from 12:00 to 5:00.)

Example: 3:23

ONLY TIME WILL TELL

1. A clock runs five minutes slow every hour. Twelve hours ago, the clock was set with the correct time. The correct time now is 3 P.M., but how many minutes will pass before the clock shows 3 P.M.?

2. At which times do the digits on a digital watch add to 3?
   (**Hint:** There are 12.)

Example: 11:01 _____   _____   _____

_____   _____   _____

_____   _____   _____

_____   _____   _____

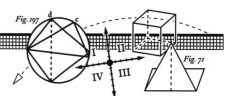

everyday

algebra

MATHEMATICAL WONDERS

Place the letters **A, B, C, D,** and **E** in the chart below so that no letter appears twice in any row, column, or diagonal.

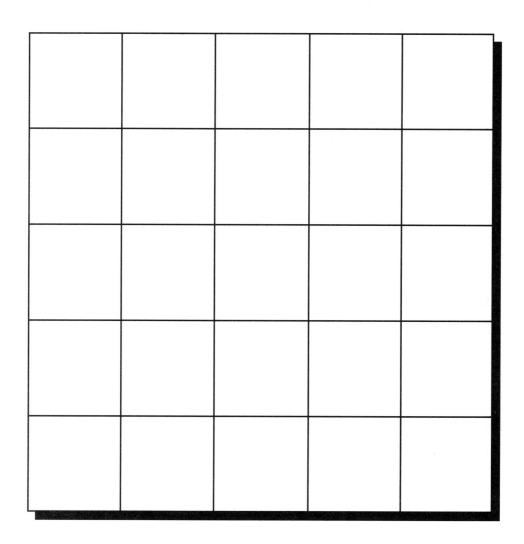

everyday **algebra**

WILL FACTOR FOR POINTS

The game below is for two players. (It can also be modified for two teams—played on the board or overhead.)

1	2	3	4	5	6
7	8	9	10	11	12
13	14	15	16	17	18
19	20	21	22	23	24
25	26	27	28	29	30
31	32	33	34	35	36

How to play:

Player A picks a number, crosses it out, and receives that number of points (for example, 20). Player B factors that number (4×5) and receives points totaling the sum of the numbers ($4 + 5 = 9$). Player B crosses out all the factors chosen. Then Player B picks a number, and the process is repeated. The game is over when all the numbers are crossed out. The player with the most points wins.

Score	
Player A	Player B

Name _____   Date _____

MISSING DIGITS—A

Find the missing digits in the problems below.

1.
```
    5 , 7 3 _
  + _ , 6 _ 5
  _____
  1 2 , _ 0 0
```

2.
```
    _ . 2 _ 7
  + 0 . 0 2 _
  _____
    7 . _ 1 8
```

3.
```
    6 . _ _
  - _ . 2 1
  _____
    2 . 1 4
```

4.
```
    _ 3 . 1 _
  -  7 . _ 7
  _____
    1 _ . 6 6
```

5.
```
    2 . _ 8 7
  + 5 . 3 _ 5
  _____
    _ . 3 7 _
```

6.
```
    8 , 9 _ 3
  - _ , 6 5 _
  _____
    5 , _ 4 8
```

everyday **algebra**

WATCH YOUR FIGURE

A. The figure below consists of 16 toothpicks of equal length. Remove 4 of the 16 toothpicks to make 4 equal triangles.

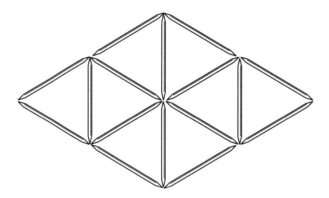

B. The figure below consists of 16 toothpicks of equal length. Given 8 additional toothpicks, divide the area into 4 equal and similar-shaped areas without moving any of the existing toothpicks.

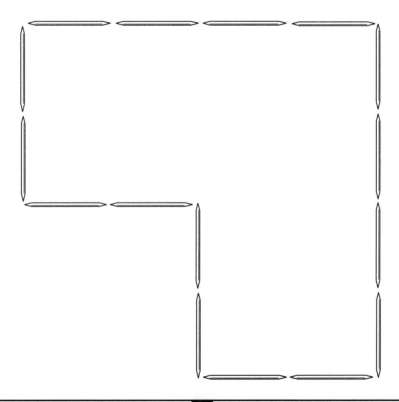

NOT NECESSARILY MATH

A. Place the letters below in the squares so that all rows and columns spell common 3-letter words. (row—left to right; column—top to bottom)
(**Hint:** Try C in the center.)

D C G E A T I E Y

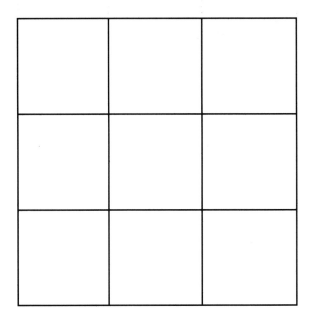

B. Three cards from a standard deck of playing cards are placed facedown on a desk. From the given clues, identify each card.

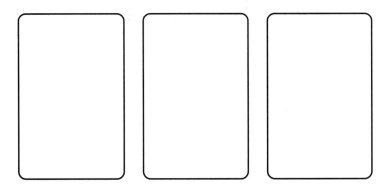

1. A jack is to the left of a heart.

2. A heart is to the left of a club.

3. A diamond is to the left of a six.

4. One card is an ace.

everyday algebra

Name_____   Date _____

O-Cubed Directions

This is a game involving order of operations. Cut out the cards on
page 24 to use in the game below.

Game of Partners: Five number cards are dealt to each player. Then a card is dealt faceup on the
desk between the two players. Players race to make their five number cards equal the number of
the card on the desk by inserting grouping symbols and any of the four operations ($+$, $-$, $\times$, $\div$).
Paper and pencil may be used.

Example: Five cards dealt to a player   card dealt faceup   the solution

1, 5, 3, 6, 10   5 $10 - (6 + 3 + 1) + 5$

Game of Many: Five cards are dealt faceup on the desk (or written on the board). Teams of up
to four students race to make all five number cards equal the numbers 1–10, using grouping
symbols and the four operations.

Example: Five cards dealt faceup   one solution

11, 9, 14, 3, 19 $11 - [(19 + 9) \div 14 + 3] = 6$

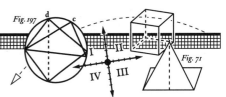

O-CUBED GAME PIECES

1	**2**	**3**	**4**	**5**
6	**7**	**8**	**9**	**10**
11	**12**	**13**	**14**	**15**
16	**17**	**18**	**19**	**20**
21	**22**	**23**	**24**	**25**

FS-10606 Everyday Algebra

CREATIVE MEASURING

1. Sally needs to brush her teeth for exactly 7 minutes—but she only has a 5-minute hourglass and a 3-minute hourglass. How can she time exactly 7 minutes using these two timers?

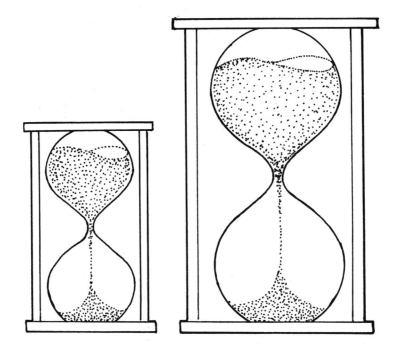

2. Chef Louis needs 2 liters of bouillon for his soup, but he only has a 5-liter container and an 8-liter container. How can he measure exactly 2 liters of the liquid to finish his soup?

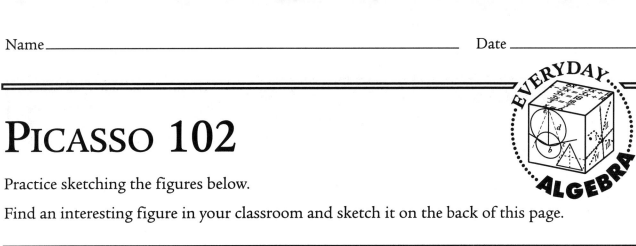

Picasso 102

Practice sketching the figures below.

Find an interesting figure in your classroom and sketch it on the back of this page.

1. stairs	**2.** a cylinder
3. an open box	**4.** a star

everyday algebra

BASKETBALL FOR SQUARES

The math club president was watching a basketball practice at her school. The players were wearing jerseys numbered 1–18. When the team began a pairs drill, the president noticed that the sum of the numbers worn by each pair of athletes was a perfect square. What were the numbers worn by the two players in each pair?

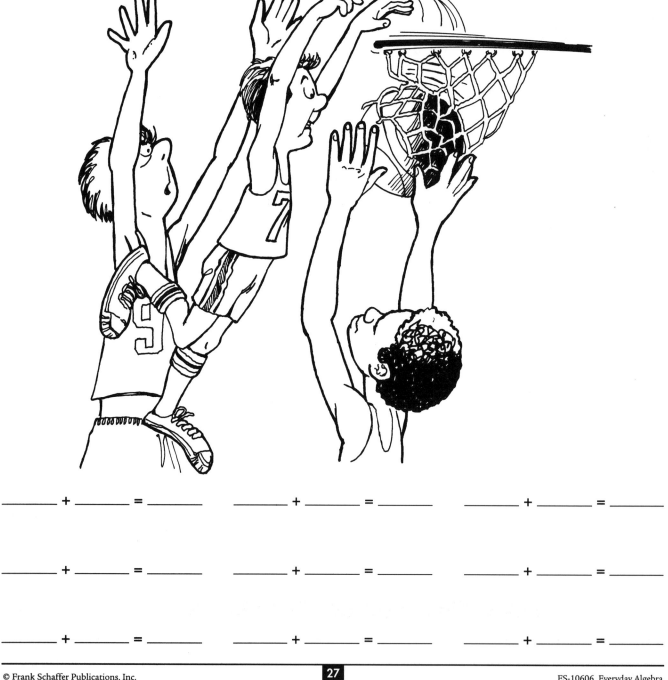

_____ + _____ = _____   _____ + _____ = _____   _____ + _____ = _____

_____ + _____ = _____   _____ + _____ = _____   _____ + _____ = _____

_____ + _____ = _____   _____ + _____ = _____   _____ + _____ = _____

FS-10606 Everyday Algebra

DIFFICULT DOMINOES

The dominoes below have been placed in the square grid in the pattern shown. Figure out where each of the 28 dominoes is, and draw lines to show its arrangement.

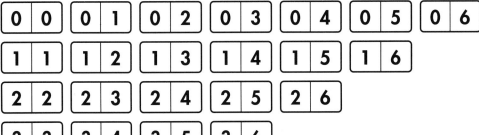

0 0	0 1	0 2	0 3	0 4	0 5	0 6
1 1	1 2	1 3	1 4	1 5	1 6
2 2	2 3	2 4	2 5	2 6
3 3	3 4	3 5	3 6
4 4	4 5	4 6
5 5	5 6
6 6

2	2	5	2	1	5	5	5
4	3	4	4	6	6	1	3
0	0	5	0	2	6	6	6
1	1	2	4	1	0	4	5
3	3	6	0	3	3	3	3
0	6	6	0	1	4	0	2
5	2	4	4	1	1	5	2

     FS-10606 Everyday Algebra

EVERYDAY

algebra

CHILD'S PLAY

Molly Math has a new set of alphabet blocks. She takes four of the blocks and notices that by arranging them differently, she can spell various words. Each of her four blocks has a different letter of the alphabet on each of its six sides. Look at the words below. Then figure out how the 24 letters are arranged on Molly's four blocks so that she is able to spell these words using her blocks.

BOXY

BUCK

CHAW

DIGS

EXAM

FLIT

GIRL

JUMP

OGRE

OKAY

PAWN

ZEST

Block 1 Letters	Block 2 Letters	Block 3 Letters	Block 4 Letters

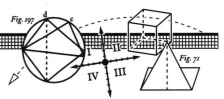

everyday

algebra

PUZZLING CIRCLE

Find the value of variables *a–j*, if . . .

1. each number 1–10 is used only once,

2. the sum of any two adjacent numbers must be equal to the sum of the two numbers in opposite sectors (i.e., $a + b = f + g$).

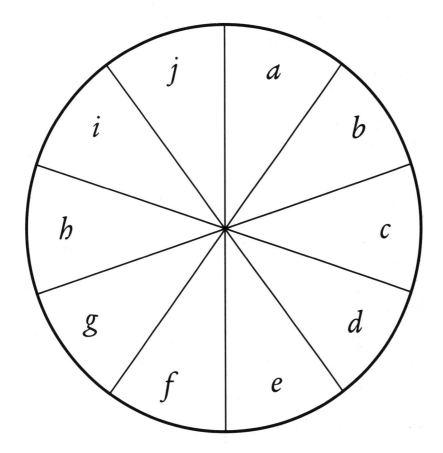

$a =$ _____ $f =$ _____

$b =$ _____ $g =$ _____

$c =$ _____ $h =$ _____

$d =$ _____ $i =$ _____

$e =$ _____ $j =$ _____

everyday **algebra**

PUZZLING NUMBERS

Find the value of variables *a–h*, if . . .

1. each number 1–8 is used only once,

2. *e* is greater than *f*,

3. *d* is less than *c*,

4. no two consecutive numbers can be in circles connected by lines.

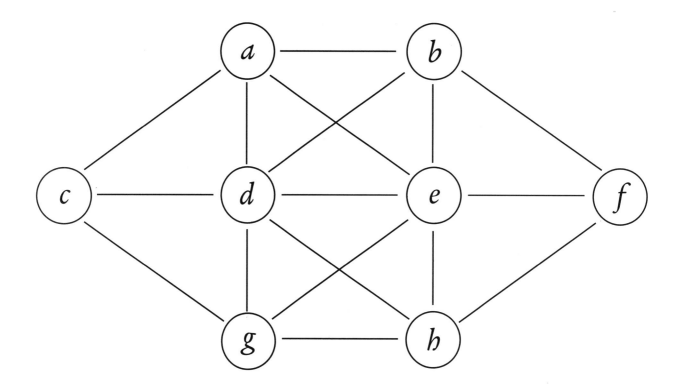

a = _____   *c* = _____   *e* = _____   *g* = _____

b = _____   *d* = _____   *f* = _____   *h* = _____

FS-10606 Everyday Algebra

NINE DOTS

1.

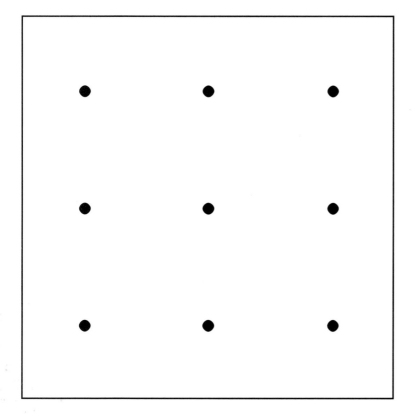

Draw four lines that cross every dot. *But . . .*

1. don't cross a dot more than once.

2. don't lift the pencil from the paper until all nine dots are crossed.

3. don't retrace a line.

2.

Draw two more squares to divide the large square into nine sections with each section containing one dot. The squares cannot cross any dots.

e v e r y d a y algebra

THE GRADUATION PARTY

Graduation season has hit Little Middle School. There are seven friends and so many parties that they need help figuring out who did what!

During one week, there was a graduation party every day. No two graduates were invited to the same party. Using the clues below, find out the day that each of the seven graduates attended a party.

1. LeAnn and Paul didn't go to a Friday or Saturday graduation party.

2. Pamela went to a Saturday party.

3. LeAnn went to a party the day before Paul did.

4. Jenny went to a party on Wednesday.

5. Paul and Alexandra didn't go on a Tuesday, but Susie did.

6. Jamie went to a party the day after Jenny.

	Sunday	Monday	Tuesday	Wednesday	Thursday	Friday	Saturday
LeAnn							
Paul							
Alexandra							
Jenny							
Jamie							
Susie							
Pamela							

QUESTIONS FOR ALL...

everyday algebra

1. All-Star Get-Together

A party was held for a group of athletes from Main Middle School. Ten of the people were on the volleyball team, 10 were on the tennis team, and 9 were on the track team. Three of the track team members were on the volleyball team, 4 of the volleyball team members were on the tennis team, and 3 of the tennis team members were on the track team. No one was on more than two teams. How many athletes were at the party?

2. All in the Family

A boy has as many sisters as brothers. Each sister has only half as many sisters as brothers. How many sisters and brothers are there?

PRIME TIME

1. One path below is paved with only prime numbers. It's the only path that will lead you to the finish line in "Prime Time!" Shade in the correct path to find the way to the finish.

Remember, the definition of a prime number is

_____

_____ .

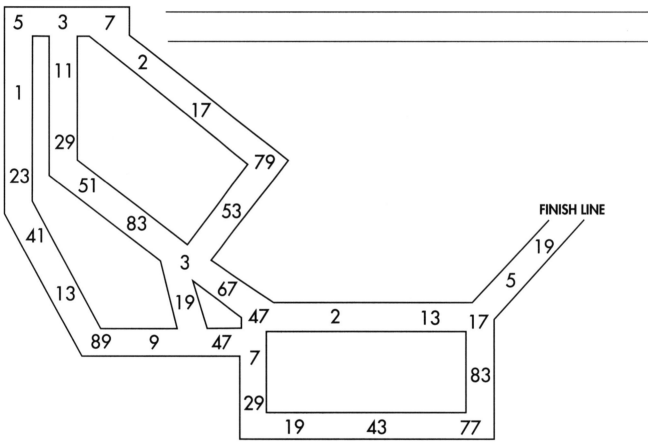

2. List all of the prime numbers less than 100.
Find four or more pairs of prime numbers with a sum of 84.

Prime numbers: _____

_____

Pairs = 84: _____

_____

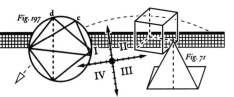

everyday

algebra

LETTER PATTERN

Find the next term in each pattern. Then explain your answer.

1. ABCDEFGH, HBCDEFGA, HGCDEFBA, _____

_____

_____

_____

_____

2. O T T F F S S E _____

_____

_____

_____

_____

3.

C	D
B	E
A	F

,

D	E
C	F
B	G

,

E	F
D	G
C	H

,

_____

_____

_____

_____

everyday **algebra**

MISSING DIGITS—B

Find the missing digits in the problems below.

1.
```
  _ 5 , 6 _ 2 , _ 1 1
+ 3 8 , _ 6 9 , 0 8 _
─────────────────────
  9 _ , 0 2 _ , 9 _ 0
```

2.
```
      4 2
  ×    _
  ───────
    2 1 0
```

3.
```
    _ 3 4
  ×   _ 2
  ───────
    1 4 _ 8
    7 _ 4 0
  ─────────
    _ 8 0 _
```

4.
```
        8 _ 8
    ×   5 1 _
  ───────────
      _ 7 9 6
      _ 9 _ 0
    4 _ 9 0 _ 0
  ─────────────
    _ 5 9 _ 7 _
```

5.
```
  9 _ , 0 3 2 , 9 _ 3
− _ 5 , 3 _ 2 , _ 9 8
─────────────────────
  7 2 , _ 4 _ , 8 1 _
```

THE FOURS HAVE IT

Using four 4s and any mathematical operations ($+$, $-$, $\times$, $\div$, $\sqrt{}$, exponents, etc.), list the numbers 0–51. This may be a long process, so hang on to this sheet and add to it often.

Example: $44 \div 4 - 4 = 7$

0 _____	26 _____
1 _____	27 _____
2 _____	28 _____
3 _____	29 _____
4 _____	30 _____
5 _____	31 _____
6 _____	32 _____
7 _____	33 _____
8 _____	34 _____
9 _____	35 _____
10 _____	36 _____
11 _____	37 _____
12 _____	38 _____
13 _____	39 _____
14 _____	40 _____
15 _____	41 _____
16 _____	42 _____
17 _____	43 _____
18 _____	44 _____
19 _____	45 _____
20 _____	46 _____
21 _____	47 _____
22 _____	48 _____
23 _____	49 _____
24 _____	50 _____
25 _____	51 _____

Name_____ Date _____

PRIME TRIANGLES

To express a number using prime factorization, divide the number by the smallest prime factor and repeat until all factors listed are prime.

Example: $36 = 2 \times 18$

$= 2 \times 2 \times 9$

$= 2 \times 2 \times 3 \times 3$

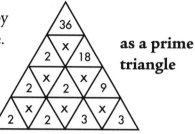

as a prime triangle

Now try a few! _____

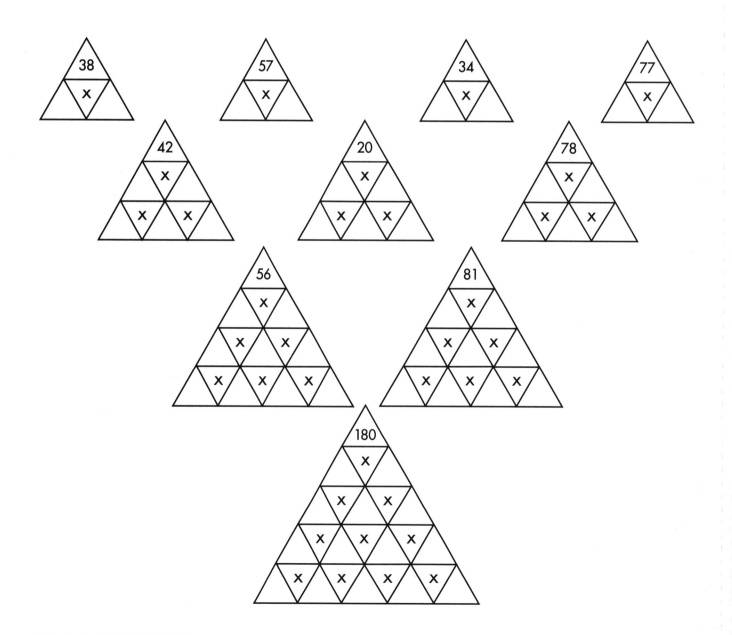

TRIANGLE OF THE 20S

Cut out the circle numbers below. Use the circles to help you correctly place the numbers in the triangle such that the sum along each side is 20. Do it again. This time, the sum along each side should be 21.

Answer–20

Answer–21

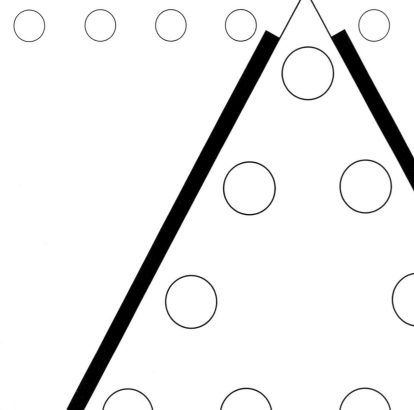

1 2 3 4 5 6 7 8 9

FS-10606 Everyday Algebra

algebra

TANGRAMS

1. Cut out the seven pieces below. Use these seven tangram pieces to make
   a. a square;
   b. a right triangle;
   c. a parallelogram;
   d. a trapezoid.

2. Use the five small pieces to form a square.

Be creative and make a sailboat, a house, a rocket, a fish, a cat, etc.

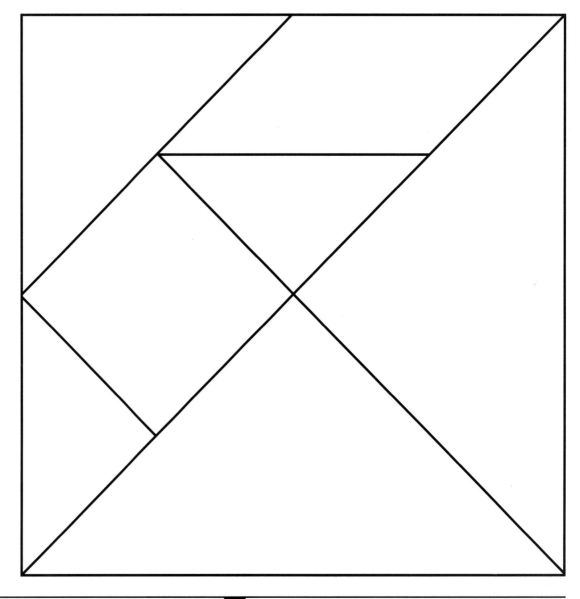

FS-10606 Everyday Algebra

everyday

algebra

ODD ONE OUT

This game is played with partners. Players take turns crossing out marks until the object of the game is reached—to leave the other person with one mark. A player may erase as many marks in one row as he or she likes during his or her turn. Play a few times and think about things such as, "Does it matter who goes first?" or, "Does it matter if you erase an odd or an even number of marks?" See if you can find a pattern.

Game 1

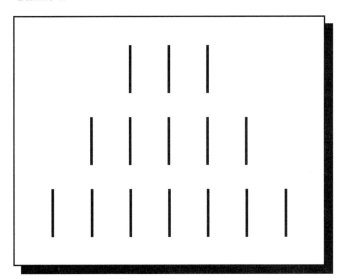

Game 2

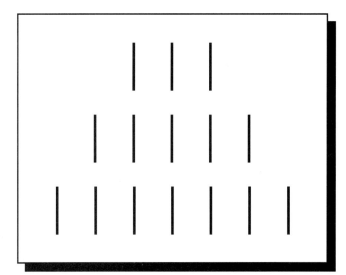

Game 3

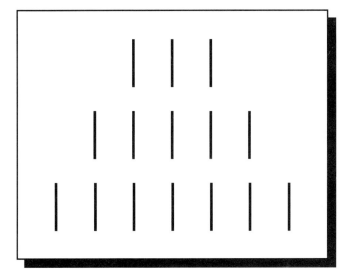

Game 4

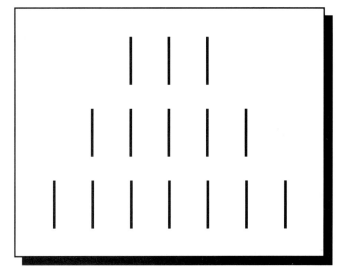

FS-10606 Everyday Algebra

Name_____ Date _____

everyday **algebra**

EQUATION SOLVING PUZZLER

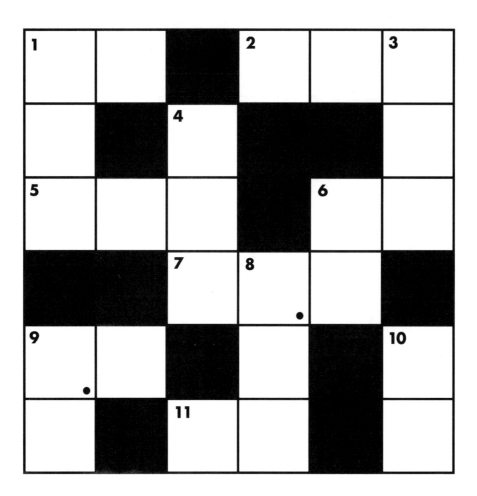

Across

1. $A + 14 = 49$

2. $15A = 4410$

5. $x + 328 = 930$

6. $\frac{2}{3}x = 40$

7. what $+ 9.8 = 33.2$?

9. 0.03 times what equals 0.072?

11. $y^2 = 324$

Down

1. $A - 98 = 258$

3. B divided by 4 equals 105.

4. $A - 296\frac{3}{4} = 25\frac{1}{4}$

6. $\frac{x}{\frac{4}{5}} = 80$

8. $x - 2.4 = 0.68$

9. $y \cdot \frac{1}{3} = 0.9$

10. $3x + 12 = 120 + 9$

 FS-10606 Everyday Algebra

AROUND THE WORLD

The sum of the numbers in two circles must equal the number in the square joining them. Find A, B, C, and D. Is there more than one solution?

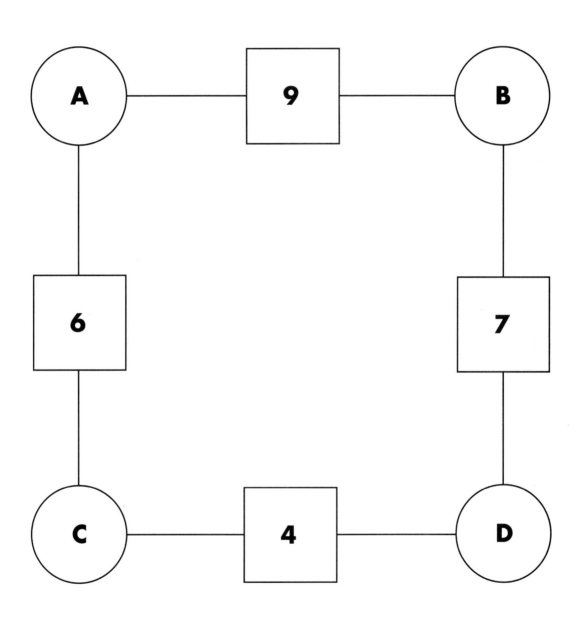

A = _____   B = _____   C = _____   D = _____

FS-10606 Everyday Algebra

WHAT SHAPE IS THAT NUMBER?

A. Triangular Numbers

are numbers that can be represented by dots that form equilateral triangles.

Examples:

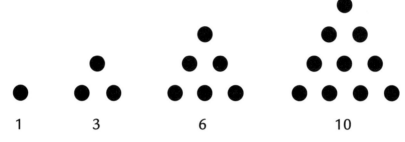

Find the next four triangular numbers.

B. Square Numbers

are numbers that can be represented by dots that form squares.

Examples:

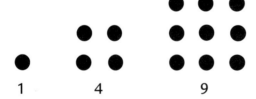

Find the next four square numbers.

 FS-10606 Everyday Algebra

PRIME BY ELIMINATION

Eratosthenes (276–195 B.C.) developed this method of finding prime numbers by "sifting out" the primes. To find all the prime numbers up to 100 (this can be done with any number), circle the 2 and cross out all the numbers that are multiples of 2. Circle the next number (3) and cross out all the numbers that are multiples of 3. Repeat this process for all numbers until only circled numbers remain. These are the prime numbers in this set!

	2	3	4	5	6	7	8	9	10
11	12	13	14	15	16	17	18	19	20
21	22	23	24	25	26	27	28	29	30
31	32	33	34	35	36	37	38	39	40
41	42	43	44	45	46	47	48	49	50
51	52	53	54	55	56	57	58	59	60
61	62	63	64	65	66	67	68	69	70
71	72	73	74	75	76	77	78	79	80
81	82	83	84	85	86	87	88	89	90
91	92	93	94	95	96	97	98	99	100

Prime numbers up to 100: _____

_____

_____

_____

everyday algebra

HOW DID THE CHICKEN CROSS THE ROAD?

This game is played with two people. Each player takes a turn coloring one line segment at a time in an effort to cross to the other side. The game should be played on a grid. Start with a 2 × 2 grid, then go to a 3 × 3 grid, and then a 4 × 4 grid, etc. The players do not have to start on an "outside" line segment. Turns may also be used to "block" your opponent.

Example:

Player A tries to get

A ↕ A

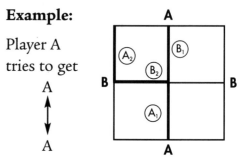

Player B tries to get B ◄──► B

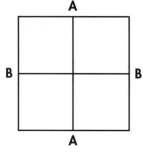

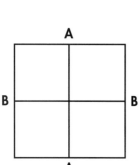

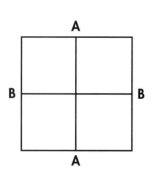

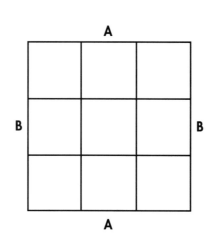

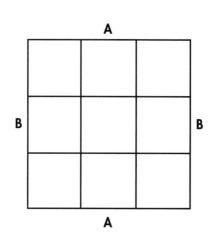

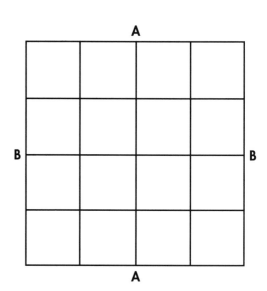

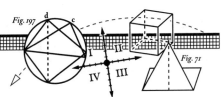

everyday

algebra

PICTURE THIS

1. How can these six sections of chain (each containing four links) be joined into one chain by cutting less than five links?

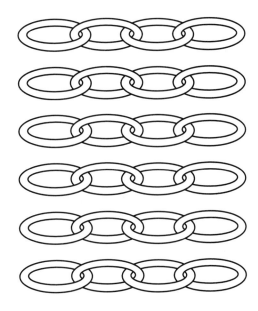

2. How can these six toothpicks be glued together to form four equilateral triangles?
   (**Hint:** Don't be so one-dimensional!)

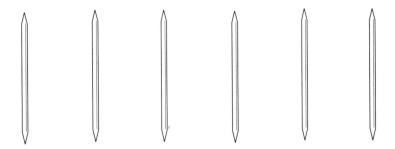

FS-10606 Everyday Algebra

Name _____ Date _____

everyday **algebra**

SUSIE SKIER—BEGINNER

Practice is important with any skill that needs improvement. Practice helps with accuracy as well as speed. Solve the problems in the ski slope below.

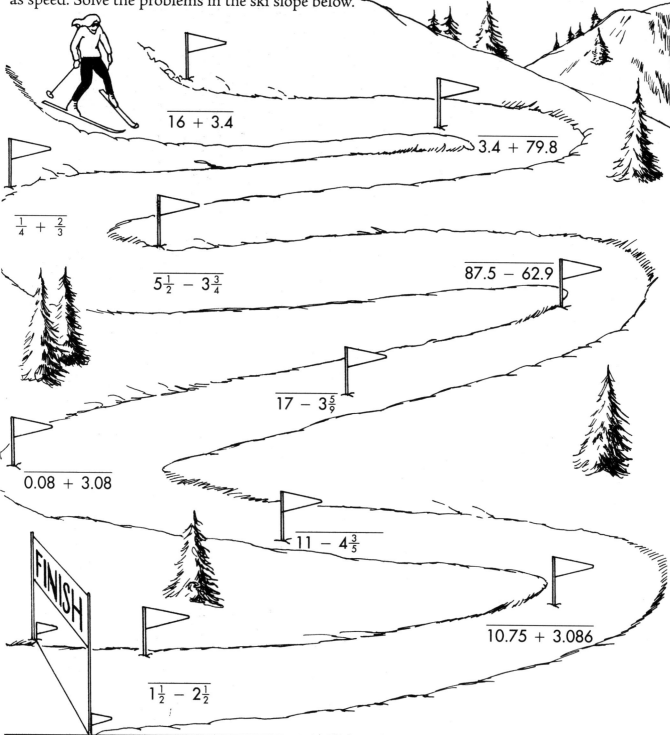

$\overline{16 + 3.4}$

$\overline{3.4 + 79.8}$

$\overline{\frac{1}{4} + \frac{2}{3}}$

$\overline{5\frac{1}{2} - 3\frac{3}{4}}$

$\overline{87.5 - 62.9}$

$\overline{17 - 3\frac{5}{9}}$

$\overline{0.08 + 3.08}$

$\overline{11 - 4\frac{3}{5}}$

FINISH

$\overline{10.75 + 3.086}$

$\overline{1\frac{1}{2} - 2\frac{1}{2}}$

49

THE GAME OF 50

Number of players: two

Who wins: The first player to reach exactly 50

Directions: Two players take turns choosing a number from 1–6. As each new number is selected, it is added to the sum of the previously selected numbers. For example, if the first student picks 2, and the second student picks 3, the sum is now 5 (second student number). The first student then picks 6, so the total is now 11 (first student number). The winner is the student who ends up with 50.

Try it, and try to find a pattern!

e v e r y d a y a l g e b r a

TRAPEZOID VOID

Formulas for Areas:

square $= s \cdot s$ or s^2

rectangle $= b \cdot h$

triangle $= \frac{1}{2} b \cdot h$

trapezoid $= \frac{1}{2}(b_1 + b_2) \cdot h$

Find the area of the shaded portion. Organize and label your work.

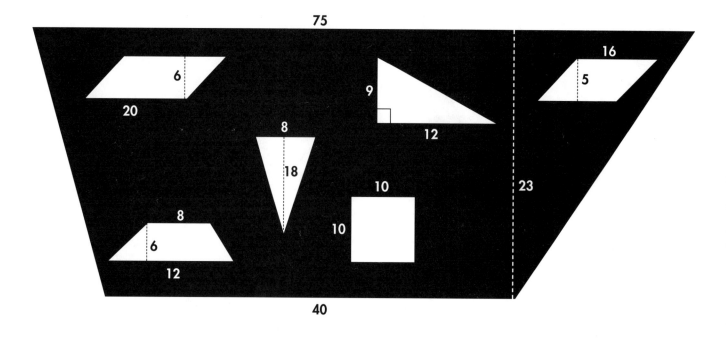

FIGURE THIS (TRIANGLE)

A box has a length of 28 cm, a width of 21 cm, and a height of 12 cm.
How long is the diagonal of the box? Show all work and logic.
(**Hint:** Use the Pythagorean theorem twice.)

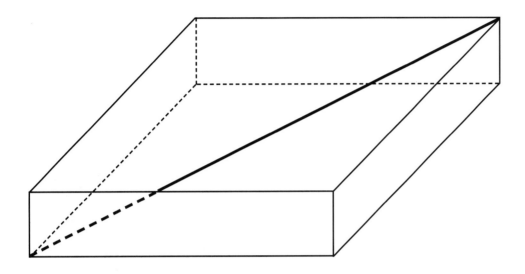

TYPICAL TRIANGLES

1. An 18-foot telephone pole is broken by a bolt of lightning, and the top falls (tips) over to the ground. If the top touches the ground 6 feet from its base, what are the lengths of the two segments of the broken telephone pole?

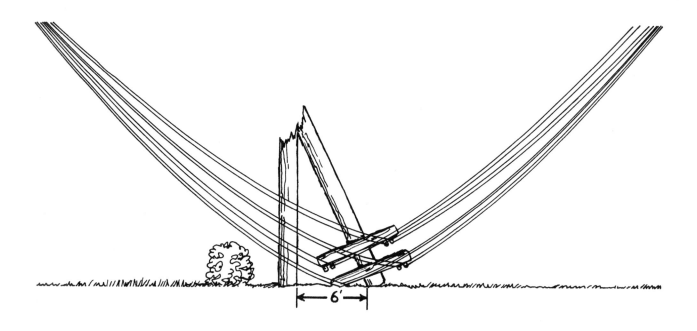

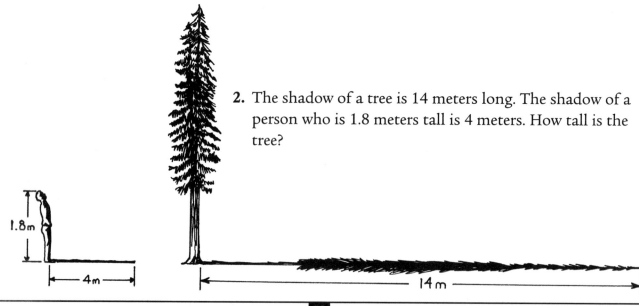

2. The shadow of a tree is 14 meters long. The shadow of a person who is 1.8 meters tall is 4 meters. How tall is the tree?

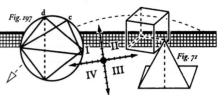

everyday

algebra

PICTURE THIS AGAIN

1. Which 3 toothpicks should be removed from the 13 in the figure below so that the remaining toothpicks will form 3 triangles?

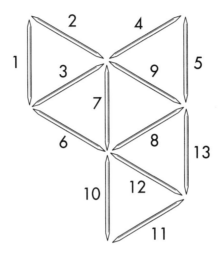

2. Move 2 toothpicks in the figure below so that a glass of the same size is formed and the penny (unmoved) is outside the glass.

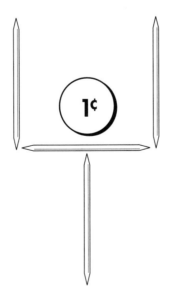

everyday **algebra**

SUSIE SKIER—INTERMEDIATE

You've mastered the bunny hill. Let's try steeper slopes.

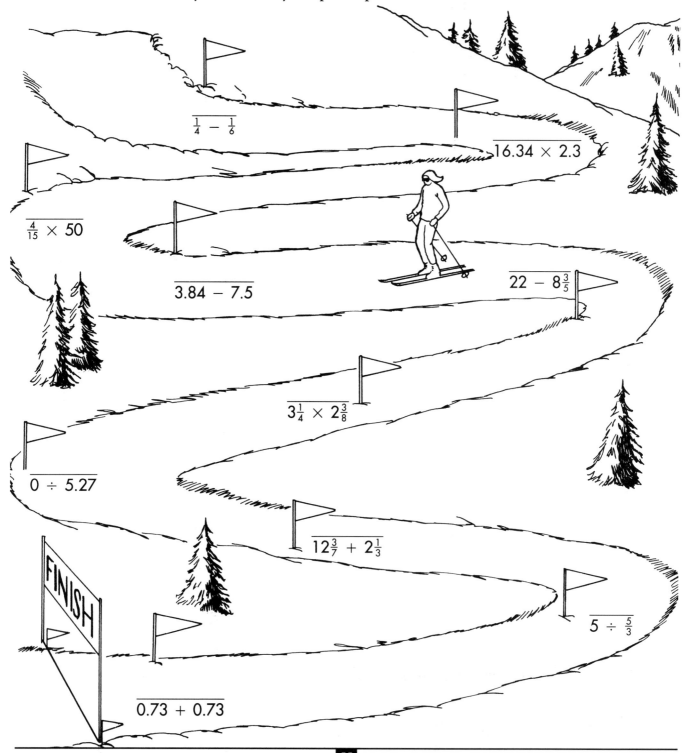

$\dfrac{1}{4} - \dfrac{1}{6}$

16.34×2.3

$\dfrac{4}{15} \times 50$

$3.84 - 7.5$

$22 - 8\dfrac{3}{5}$

$3\dfrac{1}{4} \times 2\dfrac{3}{8}$

$0 \div 5.27$

$12\dfrac{3}{7} + 2\dfrac{1}{3}$

$5 \div \dfrac{5}{3}$

$0.73 + 0.73$

FINISH

FS-10606 Everyday Algebra

FIGURE THIS (TRAPEZOID)

The longer base of a trapezoid is the square of the shorter base. The non-parallel sides are congruent. Each non-parallel side is 3 more than the shorter base. If the perimeter of the trapezoid is 24, what are the lengths of each base and the sides? Show all work and logic. (**Hint:** Draw and label a picture.)

Side _____   Side _____

Short Base _____   Long Base _____

e v e r y d a y **a l g e b r a**

SPORTS SHORTS

1. Al and Bill went bowling at the "Lanes of Leisure" and had the scores below.

Game	Winning Score	Losing Score
1	132	112
2	152	122
3	150	140

Al's total score for the three games was 424. Which game(s) did Bill win? _____

2. At a local tennis tournament, a boy noticed that the number of tennis balls in a pile by the fence was a square number. If he adds 25 more tennis balls to the pile, he could divide the whole pile into three smaller piles, each of which would contain a square number of tennis balls. The digits of the numbers representing the original pile of tennis balls add up to 10.

How many tennis balls are in the original pile? _____

How many tennis balls would there be in each of the three smaller piles? _____

FS-10606 Everyday Algebra

SUMMER SCHOOL

Andy, Beth, Carl, and Diana are enrolled in summer school P.E. class. Two classes are offered—1st hour and 2nd hour. Each student plays a different sport for the school (basketball, volleyball, swimming, and track). Find out which sport each student plays and in which hour he or she is enrolled in P.E. class. (**Hint**: Use the chart below to help you organize.)

	Volleyball	Basketball	Swimming	Track	1st Hour	2nd Hour
Andy						
Beth						
Carl						
Diana						

1. Diana is in 2nd hour.

2. Only one student in 2nd hour plays a sport that involves a ball.

3. The track athlete is in 1st hour.

4. Both the basketball player and the track athlete are boys.

5. Either Beth or Carl is in 1st hour (not both).

6. Andy and the basketball player are in different classes.

ALPHABET SOUP

1. Which number does each letter represent?
   (**Hint:** Only numbers 1–9 are used.)

$$A \times BC = DEF = HK \times G$$
$$A \times EC = DBG = AK \times F$$

2. Find the value of A.

$$
\begin{array}{r}
A\ 2\ B \\
+\quad 8\ C\ 9 \\
\hline
C\ B\ D\ 2
\end{array}
$$

3. The digits 0–7 have been grouped into "top secret categories." Break the secret code and figure out where 8 and 9 belong.

A	B	C
1	2	0
4	5	3
7		6

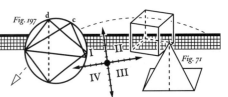

SUSIE SKIER—EXPERT

Below are the "black diamond" slopes—be cautious, but quick!

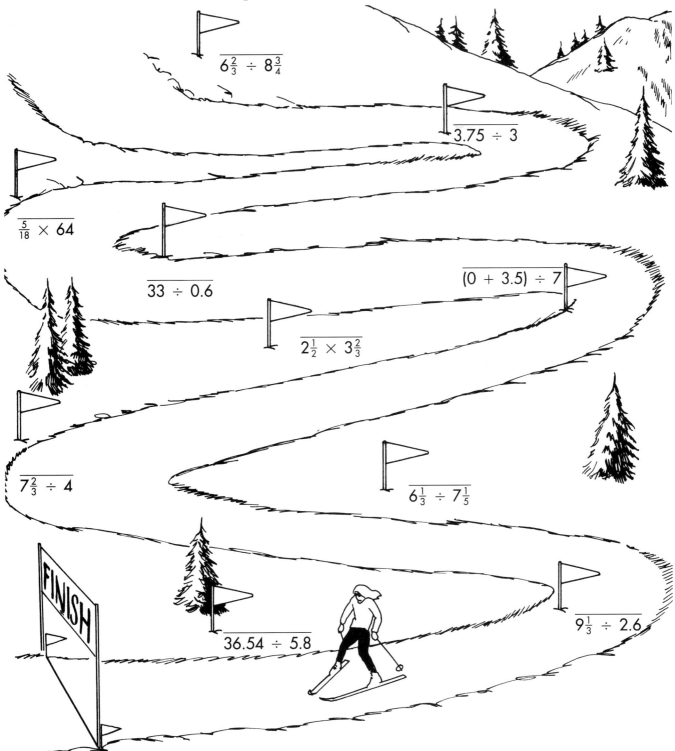

$$6\frac{2}{3} \div 8\frac{3}{4}$$

$$3.75 \div 3$$

$$\frac{5}{18} \times 64$$

$$33 \div 0.6$$

$$(0 + 3.5) \div 7$$

$$2\frac{1}{2} \times 3\frac{2}{3}$$

$$7\frac{2}{3} \div 4$$

$$6\frac{1}{3} \div 7\frac{1}{5}$$

FINISH

$$36.54 \div 5.8$$

$$9\frac{1}{3} \div 2.6$$

FS-10606 Everyday Algebra

FRACTION TRIANGLE

Cut out the circled numbers below. Use the circles to help you correctly place the numbers in the triangle such that the sum along each side of the triangle is 1.

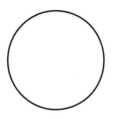

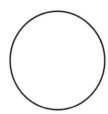
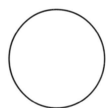

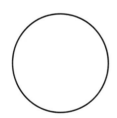
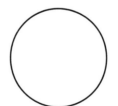
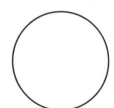

$$\frac{1}{12} \qquad \frac{5}{12} \qquad \frac{1}{2} \qquad \frac{1}{3} \qquad \frac{1}{4} \qquad \frac{1}{6}$$

FS-10606 Everyday Algebra

HOW TO GET THERE FROM HERE

Moving only vertically or horizontally, find a path from the "start block" to the "finish block" such that the sum of the numbers along the path (including start and finish) is 100.

FINISH

13	10	41	9
0	2	15	16
5	11	6	8
20	4	17	13

START

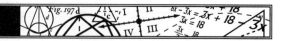

e v e r y d a y **a l g e b r a**

BUILDING BLOCKS

Find x in the figure below. Each piece is a square. A, B, and C are co-linear points.
(**Hint:** "Proportion")

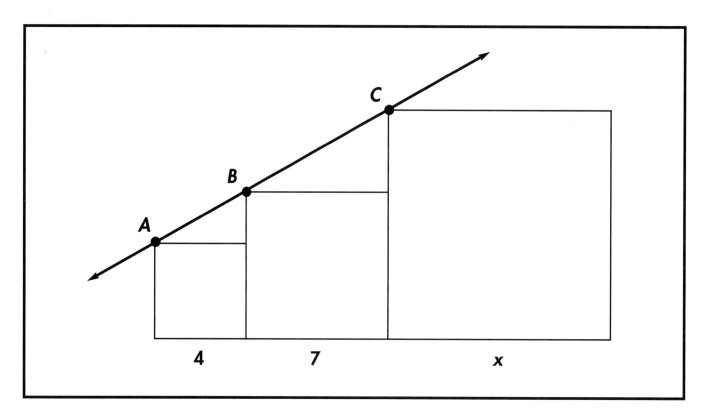

Show your work.

THERE'S A COIN IN MY TRIANGLE

A quarter is stuck (inscribed) in this triangle.
Using the given measurements, find the value of x.

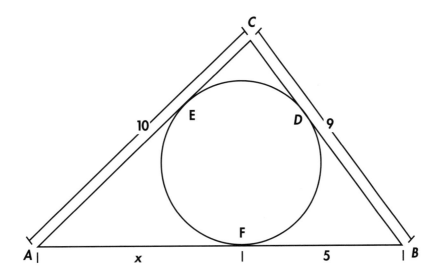

Show your work.

algebra

EVERYDAY

THE GREAT PYRAMIDS

The number in each square of each puzzle is the sum of the two numbers in the boxes above it. Find the value of each square to make the pyramids true.

1.

$x =$ _____

$y =$ _____

$z =$ _____

x		14		14
	y		z	
		73		

2.

$A =$ _____

$B =$ _____

$C =$ _____

$D =$ _____

$E =$ _____

$F =$ _____

FS-10606 Everyday Algebra

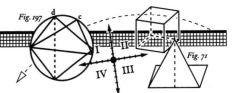

everyday

algebra

TRICKY TRIANGLES

Cut out the circled numbers below. Use the circles 1–12 for each triangle to help you correctly place the numbers in the triangle so that

a. the sum of each side is 28;

b. the sum of each side is 35.

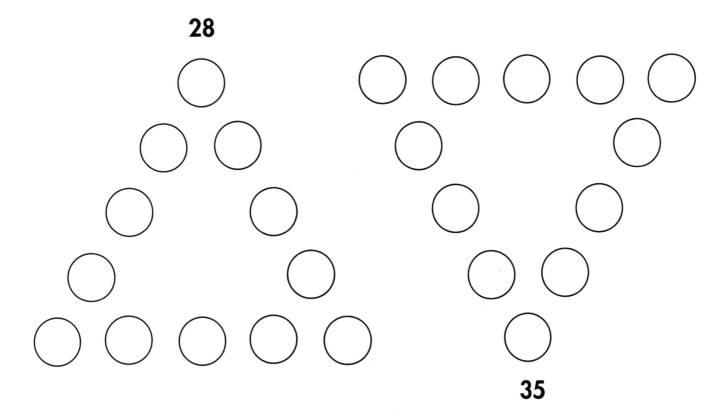

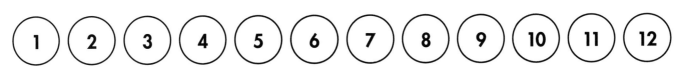

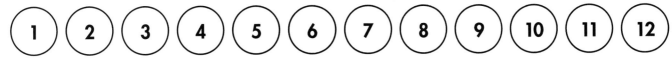

FS-10606 Everyday Algebra

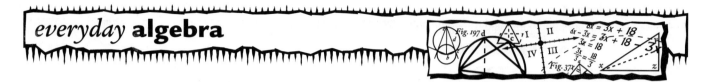

everyday **algebra**

WEIGHING WONDER

Of eight identical-looking golf balls, seven weigh the same, but one is slightly lighter than the others. Find the lighter ball only using two weighings on the scale below. Explain how you can do this.

Explanation: _____

_____

_____

_____

67

SUM/PRODUCT TRIANGLES

1. The number in each square is the sum
of the two circles next to it.
Find the missing values.

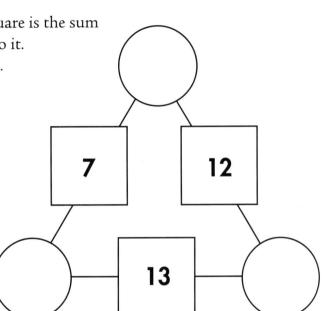

2. Cut out the integers below. Use the
circles to help you correctly place
the integers such that the product
of each side is 60.

2   4   ‾3   ‾5   ‾6   ‾10

FS-10606 Everyday Algebra

ORDER IN THE SQUARE

Cut out the square numbers below. Place the squares in the same position as shown in Figure A. By sliding only one square at a time into an open square (do not lift and move the pieces), arrange the pieces into the positions shown in Figure B.

Figure A

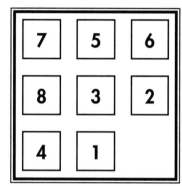

Figure B

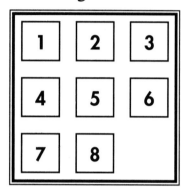

| 1 | 2 | 3 | 4 | 5 | 6 | 7 | 8 |

VISUAL COUNTING

1. How many cubes are there in the figure below?

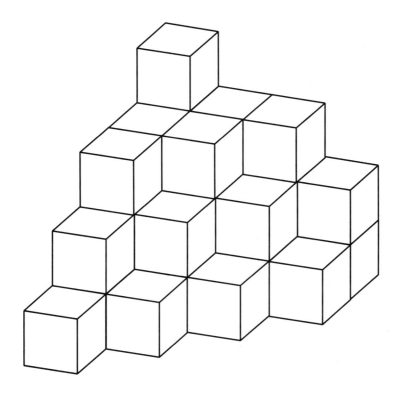

2. How many rectangles are there in the figure below?

FS-10606 Everyday Algebra

everyday algebra

$6x =$
$6x - 3x = 3$
$3x = 18$
$\dfrac{3x}{3} = \dfrac{18}{3}$
$x = 6$
Fig.

POLKA DOTS AND DICE

1. Place five dots in five squares so that no two dots are in line vertically, horizontally, or diagonally.

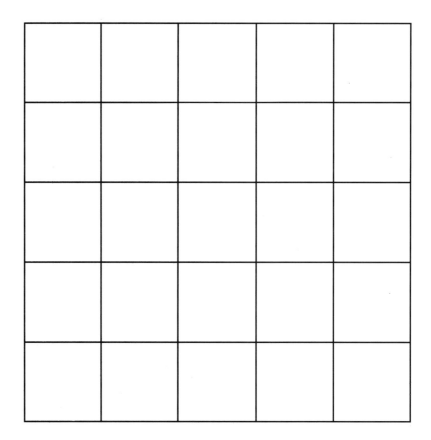

2. Three identical dice are thrown and land as shown. Figure out the exact make-up of the dice.

number opposite the

one _____

number opposite the

four_____

number opposite the

five _____

Name_____ Date_____

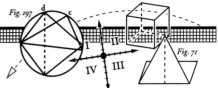

GIVE YOURSELF A BIG HAND

1. Estimate the area of your hand.

2. Trace your hand on the grid below. Using various area concepts, approximate the actual area of your hand. Be as exact as possible.

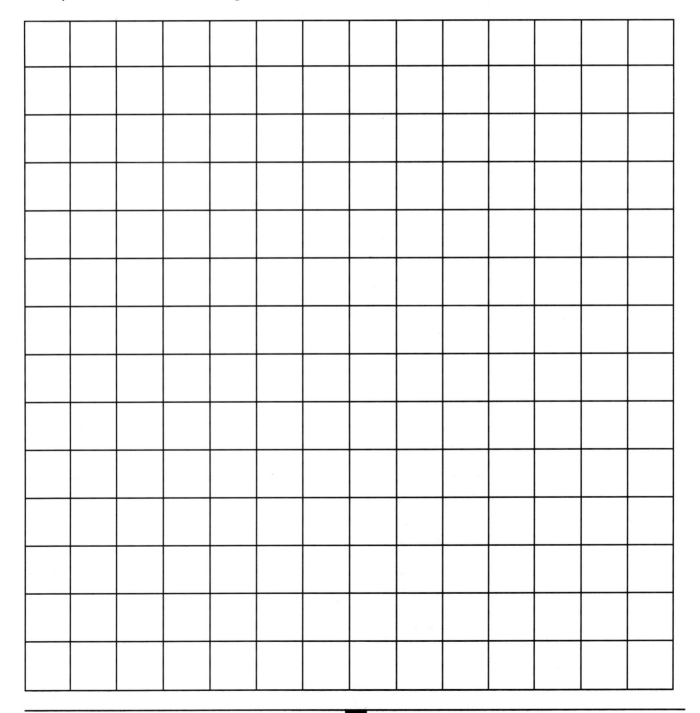

everyday **algebra**

WORTH THE WEIGHT!

Based on the two scales shown, determine how many marbles will balance one cup.

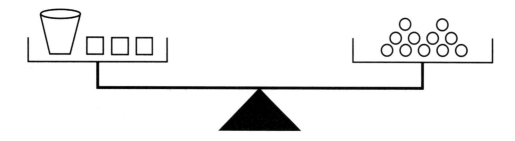

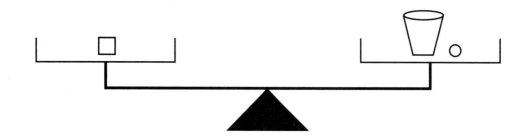

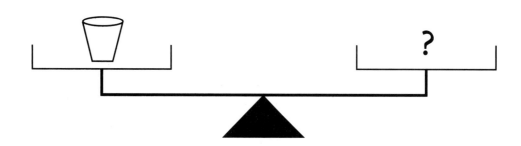

FOLLOW THE PATH

Begin in the upper left-hand corner of this puzzle. Move along every possible path diagonally, horizontally, or vertically and add up the numbers along that path. How many paths total 9?

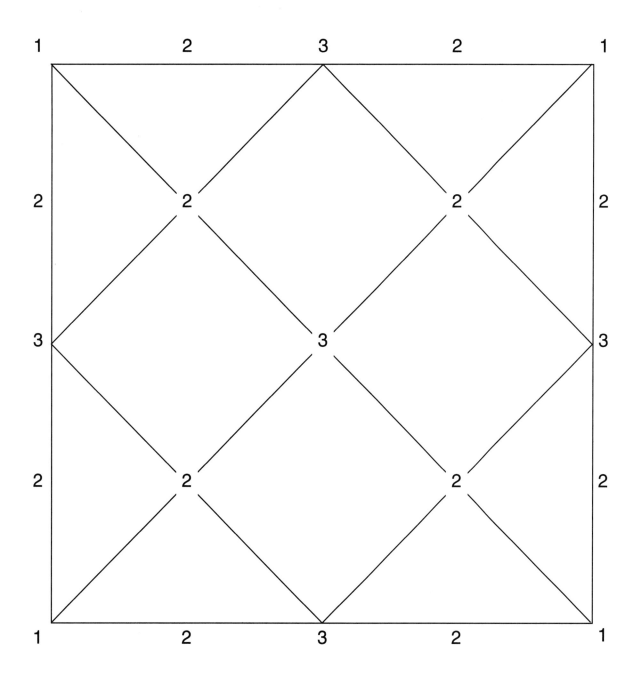

ANSWER KEY

Page 1

3	x	8	÷	4	=	6
+	■	+	■	x	■	−
6	÷	2	+	3	=	6
−	■	−	■	−	■	+
7	+	5	−	7	=	5
=	■	=	■	=	■	=
2	x	5	−	5	=	5

5	x	3	−	9	=	6
+	■	x	■	+	■	÷
4	÷	2	+	1	=	3
−	■	−	■	−	■	+
3	x	4	−	5	=	7
=	■	=	■	=	■	=
6	x	2	+	5	=	9

Page 2

1.

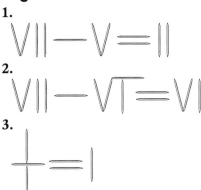

VII − V = II

2.

VII − VT = VI

3.

+ = I

Page 3

1. 8
2. a. 8; **b.** 0; **c.** 0; **d.** 0
3. (1) 27
   (2) a. 8; **b.** 12; **c.** 6; **d.** 1
4. (1) 64
   (2) a. 8; **b.** 24; **c.** 24; **d.** 8

Page 4

Anna—bedroom
Beth—living room
Christopher—bathroom
Dean—den
Ellen—kitchen

Page 5

1.

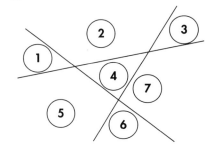

2.

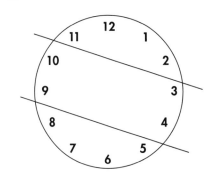

Page 6

A.

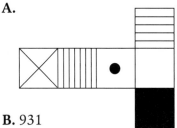

B. 931

Page 7

1. 92 chickens, 4 goats, 4 cows
2. Two possible combinations—
1 quarter, 2 dimes, 2 nickels, and
45 pennies; or 2 dimes, 8 nickels,
and 40 pennies

Page 8

Answers will vary. Possible
answers include:

$(3 \times 3) \div (3 + 3 + 3) = 1$
$(3 + 3 + 3 − 3) \div 3 = 2$
$3 + 3 − 3 + 3 − 3 = 3$
$33 \div 33 + 3 = 4$
$3 \times 3 − 3 − (3 \div 3) = 5$
$(3 \times 3) + 3 − 3 − 3 = 6$
$(33 − 3) \div 3 − 3 = 7$
$(3 + 3) \div 3 + 3 + 3 = 8$
$3 + 3 + 3 + 3 − 3 = 9$
$33 \div 3 − (3 \div 3) = 10$

Page 9

1.

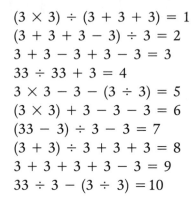

2.

Page 10

1.   **2.**

3.   **4.**

FS-10606 Everyday Algebra

Page 11
Answers will vary.

Page 12
1. yes
2. yes

Page 14
1. 8164, 1649, 3649
2. $x = 27; y = 594; z = 16{,}038$

Page 15
A. 6 chickens, 9 horses
B. 108 pounds total,
sack 1 = 22 pounds;
sack 2 = 26 pounds;
sack 3 = 28 pounds;
sack 4 = 18 pounds;
sack 5 = 14 pounds

Page 16
1. 1:11, 2:22, 3:33, 4:44, 5:55, 10:00, 11:10, 11:11, 11:12, 11:13, 11:14, 11:15, 11:16, 11:17, 11:18, 11:19, 12:22
2. 12:21, 1:01, 1:11, 1:21, 1:31, 1:41, 1:51, 2:02, 2:12, 2:22, 2:32, 2:42, 2:52, 3:03, 3:13, 3:23, 3:33, 3:43, 3:53, 4:04, 4:14, 4:24, 4:34, 4:44, 4:54

Page 17
1. 65 minutes
2. 11:01, 10:02, 10:11, 10:20, 11:10, 12:00, 1:02, 1:11, 1:20, 2:01, 2:10, 3:00

Page 18
Two possible solutions:

A	D	B	E	C
B	E	C	A	D
C	A	D	B	E
D	B	E	C	A
E	C	A	D	B

A	C	E	B	D
B	D	A	C	E
C	E	B	D	A
D	A	C	E	B
E	B	D	A	C

Page 20
1.
$$\begin{array}{r} 5{,}735 \\ +\ 6{,}665 \\ \hline 12{,}400 \end{array}$$

2.
$$\begin{array}{r} 7.297 \\ +\ 0.021 \\ \hline 7.318 \end{array}$$

3.
$$\begin{array}{r} 6.35 \\ -\ 4.21 \\ \hline 2.14 \end{array}$$

4.
$$\begin{array}{r} 23.13 \\ -\ 7.47 \\ \hline 15.66 \end{array}$$

5.
$$\begin{array}{r} 2.987 \\ +\ 5.385 \\ \hline 8.372 \end{array}$$

6.
$$\begin{array}{r} 8{,}903 \\ -\ 3{,}655 \\ \hline 5{,}248 \end{array}$$

Page 21
A.

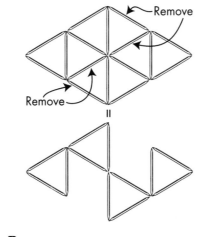

B.

Page 22
A.

D	I	G
A		C	E
Y	E	T

B.

Jack of Diamonds   Six of Hearts   Ace of Clubs

Page 25
1. Start the hourglasses together. When the 3-minute hourglass is finished, there will be 2 minutes left on the 5-minute hourglass. Begin brushing now. When the 5-minute hourglass is finished, turn it over immediately, and start the 5-minute hourglass again. (2 + 5 = 7)
2. Fill the 5-liter container, and pour it into the 8-liter container. Fill the 5-liter container again, and pour as much soup as possible (3 liters) into the 8-liter container. When the 8-liter container is full, there will be only 2 liters left in the 5-liter container.

Page 27
$18 + 7 = 25$
$17 + 8 = 25$
$16 + 9 = 25$
$15 + 1 = 16$
$14 + 2 = 16$
$13 + 3 = 16$
$12 + 4 = 16$
$11 + 5 = 16$
$10 + 6 = 16$

Page 28

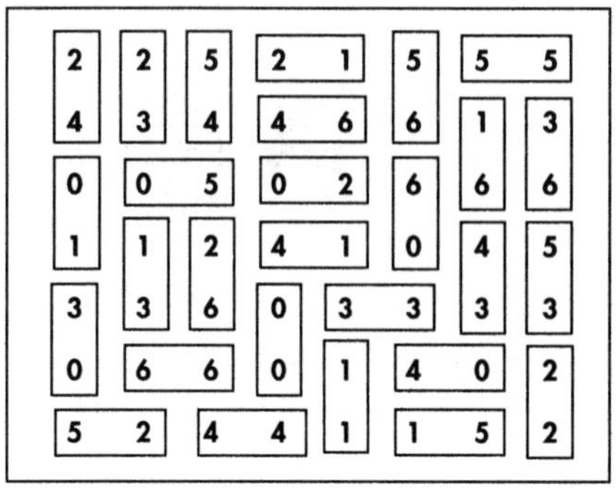

Page 29

Block 1: D, E, L, U, W, Y
Block 2: C, I, M, N, O, Z
Block 3: G, H, K, P, T, X
Block 4: A, B, F, J, R, S

Page 30

$a = 10; b = 1; c = 8; d = 3;$
$e = 6; f = 9; g = 2; h = 7; i = 4;$
$j = 5$

Page 31

$a = 3; b = 5; c = 7; d = 1;$
$e = 8; f = 2; g = 4; h = 6$

Page 32

1.

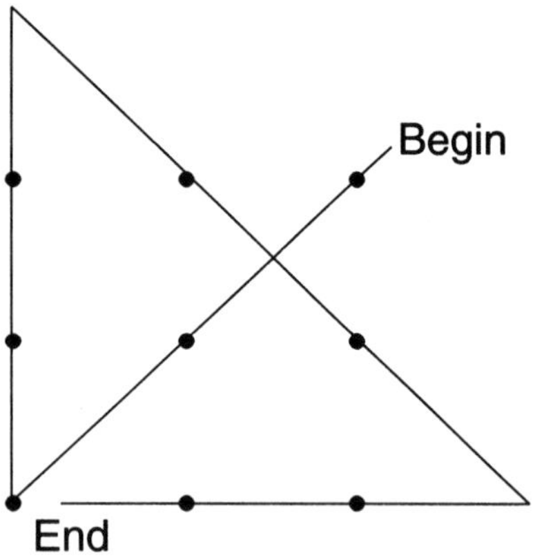

2.

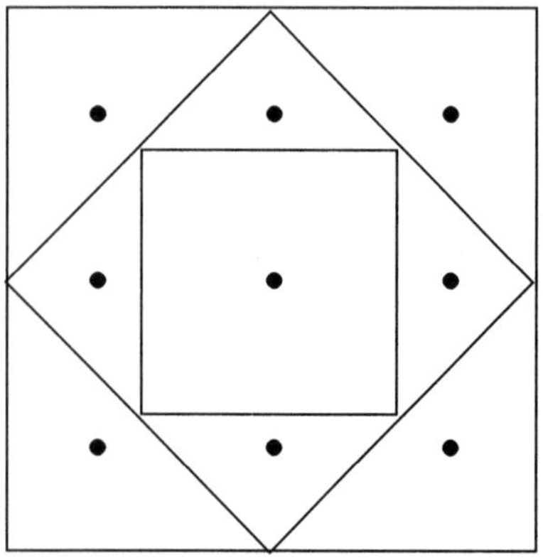

Page 33

LeAnn: Sunday
Paul: Monday
Alexandra: Friday
Jenny: Wednesday
Jamie: Thursday
Susie: Tuesday
Pamela: Saturday

Page 34

1. 19
2. 4 brothers/3 sisters

Page 35

1. A number greater than 1 whose only factors are 1 and itself
2. Prime numbers: 2, 3, 5, 7, 11, 13, 17, 19, 23, 29, 31, 37, 41, 43, 47, 53, 59, 61, 67, 71, 73, 79, 83, 89, 97

Path: 7, 2, 17, 79, 53, 3, 67, 47, 2, 13, 17, 5, 19

Pairs: $79 + 5 = 84$, $73 + 11 = 84$, $67 + 17 = 84$, $53 + 31 = 84$, $47 + 37 = 84$, $41 + 43 = 84$

Page 36

1. HGFDECBA
2. N (numbers)
3.

F	G
E	H
D	I

Page 37

1.
$$\begin{array}{r} 55{,}652{,}811 \\ +\ 38{,}369{,}089 \\ \hline 94{,}021{,}900 \end{array}$$

2.
$$\begin{array}{r} 42 \\ \times\ 5 \\ \hline 210 \end{array}$$

3.
$$\begin{array}{r} 734 \\ \times\ 12 \\ \hline 1468 \\ 7340 \\ \hline 8808 \end{array}$$

4.
$$\begin{array}{r} 898 \\ \times\ 512 \\ \hline 1796 \\ 8980 \\ 449000 \\ \hline 459776 \end{array}$$

5.
$$\begin{array}{r} 98{,}032{,}913 \\ -\ 25{,}392{,}098 \\ \hline 72{,}640{,}815 \end{array}$$

Page 38

Answers will vary.

Page 39

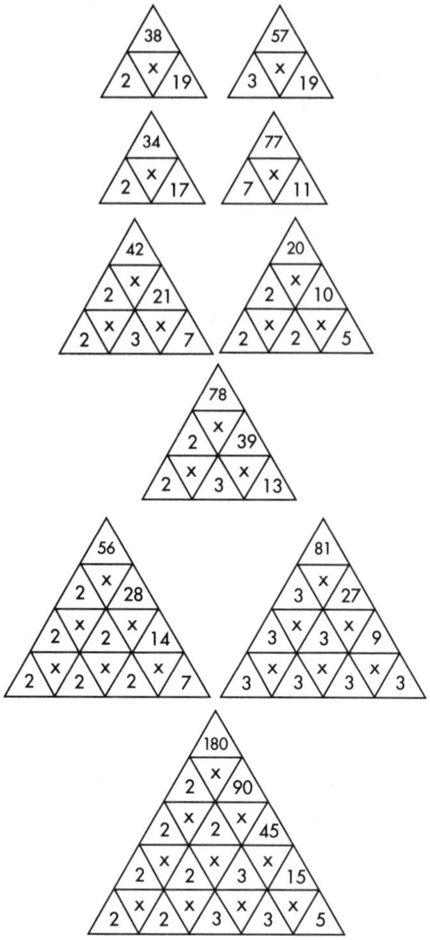

Page 40

20

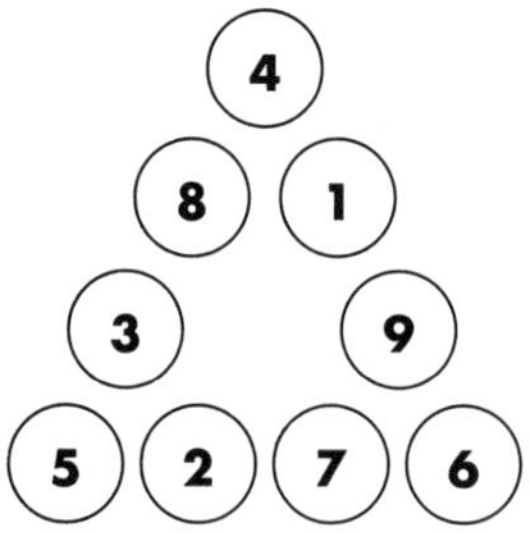

21

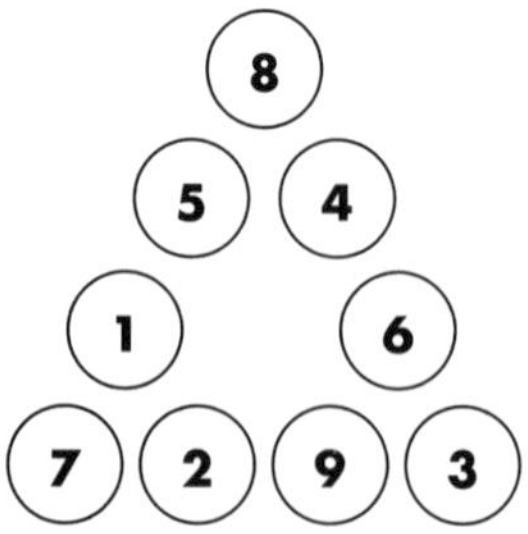

Page 41

1. a.

b.

c.

d.

2.

Page 43

¹3	5	■	²2	9	³4
5	■	⁴3	■	■	2
⁵6	0	2	■	⁶6	0
■	■	⁷2	⁸3 .	4	■
⁹2 .	4	■	0	■	¹⁰3
7	■	¹¹1	8	■	9

Page 44

$A = 5; B = 4; C = 1; D = 3$

Page 45

A. 15, 21, 28, 36
B. 16, 25, 36, 49

Page 46

2, 3, 5, 7, 11, 13, 17, 19, 23, 29, 31, 37, 41, 43, 47, 53, 59, 61, 67, 71, 73, 79, 83, 89, 97

Page 48

1. Cut bottom chain and hook all other rows together.

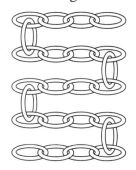

2. Form a tetrahedron

Page 49

19.4, 83.2, $\frac{11}{12}$, $1\frac{3}{4}$, 24.6, $13\frac{4}{9}$, 3.16, $6\frac{2}{5}$, 13.836, 4

Page 50

Pattern: You can always reach 50 first if you reach 43 first because regardless of what your opponent selects, you can choose one of the numbers from 1–6 to reach 50. The same logic would indicate that to win, you should obtain one of the numbers 1, 8, 15, 22, 29, 36, 43, or 50 as soon as you can, then just always pick the complement of your opponent's number in terms of "7" (i.e., if opponent picks 2, you pick 5).

Page 51

836.5 units²

Page 52

First find diagonal along bottom.

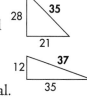

Then find diagonal.

Answer: 37 cm

Page 53

1.

$x + y = 18$
$x = 18 - y$
so $(18 - y)^2 + 6^2 = y^2$
$y = 10$
$x = 8$

2. Use ratios.
$\frac{x}{14} = \frac{1.8}{4}$
$x = 6.3$ m

Page 54

1. 7, 8, 9

2.

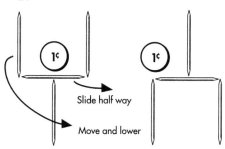

Slide half way

Move and lower

Page 55

$\frac{1}{12}$, 37.582, $13\frac{1}{3}$, -3.66, $13\frac{2}{5}$, $7\frac{23}{32}$, 0, $14\frac{16}{21}$, 3, 1.46

Page 56

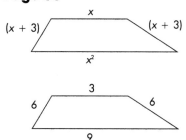

Page 57

1. Games 1 and 2
2. 1225 tennis balls, 225/400/625

FS-10606 Everyday Algebra

Page 58
Andy—track—1st
Carl—basketball—2nd
Beth—volleyball—1st
Diana—swimming—2nd

Page 59
1. $2 \times 78 = 156 = 39 \times 4$
   $3 \times 58 = 174 = 29 \times 6$

2. $A = 5$ ($B = 3, C = 1, D = 4$)

3. 8 & 9 both belong in C.
(curved numbers), $A =$ all
straight lines; $B =$ half curved
and half straight

Page 60
$\frac{16}{21}$, 1.25, $17\frac{7}{9}$, 55, 0.5, $9\frac{1}{6}$, $1\frac{11}{12}$,
$\frac{95}{108}$, $3\frac{23}{39}$, 6.3

Page 61

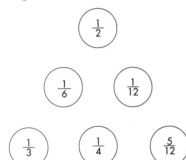

Page 62
20 ▶ 5 ▶ 0 ▶ 13 ▶ 10 ▶ 2 ▶ 11
▶ 6 ▶ 8 ▶ 16 ▶ 9

Page 63

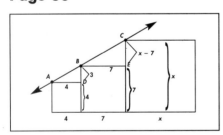

Set up ratios,

$\frac{BD}{DA} = \frac{CE}{EB}$ ▶ $\frac{3}{4} = \frac{x-7}{7}$

$21 = 4x - 28$

$x = 12\frac{1}{4}$

Page 64
If $BF = 5$, then $BD = 5$.
Therefore, $DC = 4$ and $CE = 4$.
Therefore, $EA = 6$ and $AF = 6$,
or $x = 6$.

Page 65
1. $x = 31, y = 45, z = 28$
2. $A = 20, B = 35, C = 22,$
$D = 25, E = 57, F = 47$

Page 66

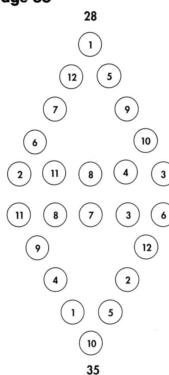

Page 67
Choose 6 of the 8 balls, and put
3 in each pan on the scale. If the
scale balances, then the lighter
ball is 1 of the 2 not chosen.
Remove the 6 balls, and take the
2 not yet weighed and put 1 in
each pan. The lighter ball will be
obvious. If the scale doesn't
balance, then the lighter ball is
in the higher pan. Remove all 6
balls and take the 3 from the
lighter side, putting 1 on each
pan and 1 not weighed. If the
scale balances, the lighter 1 is
the 1 that is out. If the scale
doesn't balance, the lighter one
will be obvious.

Page 68
1.

2.

Page 70
1. 31
2. 36 (9 individuals, 12 with
2 smaller, 6 with 3 smaller,
4 with 4 smaller, 4 with
6 smaller, and 1 with all 9)

Page 71
1.

2.
opposite the one is six
opposite the four is three
opposite the five is two

Page 73

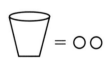

Page 74
80